Solar Low Energy Houses of IEA Task 13

January 1995

Austria

Belgium

Canada

Denmark

Finland

Germany

Japan

Netherlands

Norway

Sweden

Switzerland

United States

Solar Heating & Cooling Programme
International Energy Agency

Acknowledgements

Editor:	Robert Hastings	Solararchitektur, ETH Zurich
Working group:	Hans Erhorn	Fraunhofer Inst. für Bauphysik, Stuttgart
	Robert Hastings	Solararchitektur, ETH Zürich
	Anne Grete Hestnes	NTH, Trondheim
	Bjarne Saxhof	Thermal Insul. Lab., DTH, Lyngby

Statistics & analyses:	Douglas Balcomb	NREL, Golden, CO, USA
House information:	See individual chapters for contributors	
Design drawings:	Andreas Blaser	Solararchitektur, ETH Zurich
System diagrams:	Inger Andresen	SINTEF, Trondheim
Chapter graphs:	Michael Beckert	Fraunhofer Inst. für Bauphysik, Stuttgart

Production:	Annuscha Schmidt	Solararchitektur, ETH Zurich

Production financing:	Swiss Federal Office of Energy, Berne

Published by James & James (Science Publishers) Ltd, 5 Castle Road, London NW1 8PR

ISBN 1-873936-37-0

British Library Cataloguing in Publication Data:
A catalogue record for this book is available from the British Library.

Printed by Shiny International Ltd, Hong Kong

Table of contents

Foreword

The fifteen solar low-energy houses presented here have been built, or are being built, as part of an international collaboration within the framework of Task 13 of the IEA-SHAC[1]. The predicted total annual energy consumption for all end uses averages only 45 kWh/m². Of this, auxiliary space heating comprises only 16 kWh/m² and electricity demand only 18 kWh/m².

To achieve these extremely low levels and provide superior comfort, diverse conservation and solar strategies were used, and energy-saving appliances were specified. In some cases grid-connected photovoltaic systems meet part or even all of the electric load on an annual basis.

Because the goal of Task 13 was to encourage innovation and experiment with new technologies to drastically reduce energy consumption, cost-effectiveness was not imposed as a constraint. None-the-less, the houses had to be subsequently sold, so the costs could not be too far out of line.

This collection of houses is also interesting in that it reflects the diverse cultures, and differences in design and construction of the various countries. Extremes in climates are represented, ranging from cold Scandinavian locations to mild maritime sites. The locations of the houses are keyed on the map in figure 1 overleaf. Addresses of contact persons and information about Task 13 and the IEA Solar Heating & Cooling Programme are provided in the appendices.

---

1 Solar Heating & Cooling Programme of the International Energy Agency

5

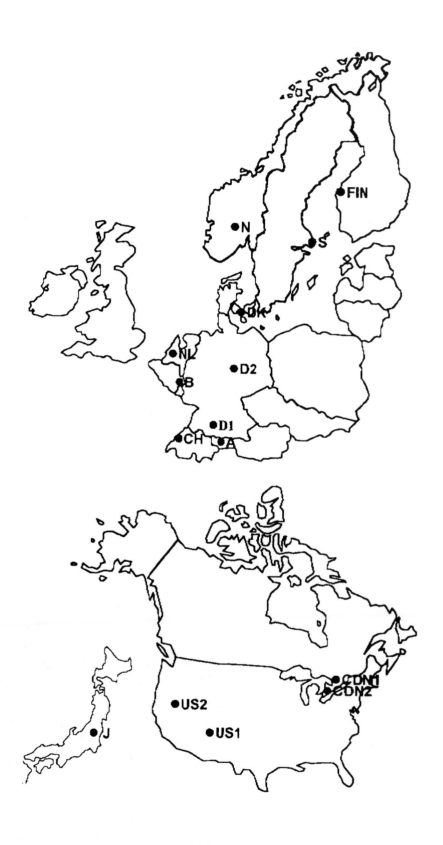

Fig.1: Locations of the IEA-Solar Low Energy Houses

An Overview of Strategies Used in the IEA Houses

Introduction

These 15 houses have been designed to achieve very low energy consumption by applying simple passive solar design principles and then, in some cases, adding "high-tech" systems. To cut losses, the building shell is typically very well insulated (U-values less than 0.2 W/m²K) and the surface-to-volume ratio is kept small. In most houses, heat is recovered from exhaust air and waste water either by a heat exchanger or a heat pump. To maximise the solar contribution diverse approaches were taken, ranging from passive direct gain through advanced windows (U-centre of glass < 1.3 W/m²K) to a collector facade with seasonal heat storage. Figure 2 provides an overview of the various strategies used in the houses.

		Super insulation	Ground coupled heat exchanger	High performance windows	Sunspaces	Solar hot water systems	Photovoltaic systems	Heat recovery systems	Integrated mechanical systems	Home automation systems	Low energy appliances	Solar control
A	Austria	X	X	X		X		X			X	
B	Belgium	X						X		X		X
CD-1	Canada	X		X	X			X	X		X	
CD-2	Canada	X	X	X		X		X	X		X	X
DK	Denmark	X		X		X		X			X	
FIN	Finland	X	X	X	X	X	X	X			X	
D-1	Germany	X		X				X			X	
D-2	Germany	X	X	X	X	X		X			X	
J	Japan				X	X	X	X				
NL	Netherland	X		X	X	X		X	X	X		
N	Norway	X	X	X	X		X	X	X		X	
S	Sweden	X		X		X		X			X	X
CH	Switzerland	X		X								
US-1	USA	X		X				X	X		X	X
US-2	USA							X	X		X	X

Fig. 2: Conservation and solar strategies of the IEA houses

The climates

Very different climates are represented in this sample of houses. This diversity is evident when temperatures and solar radiation are considered for the heating seasons (arbitrarily defined as from 1 October to 31 March). The heating degree days (base 20°C) range from c. 4600 for the Finnish house (FIN) to 1800 for the US-2 house. Solar radiation on the south facade during the heating season ranges from nearly 900 kWh/m² for the US-1 house to barely over 100 kWh/m² for the Finnish house. Quite interestingly, there is little relationship between severity of climate and heating energy use or total energy use. The reason for this is that houses in colder climates are typically better insulated. The Finnish house, confronted with the coldest climate, achieved the second lowest auxiliary heating consumption.

Space heating

The designers of these houses were successful in drastically reducing space heating demand, typically to less than a third of the total energy demand. The major strategies used are:

– Minimise the heat losses
– Profit from passive solar and internal gains
– Heat with an active solar system
– Heat with recovered heat
– Provide auxiliary heat efficiently.

Conservation measures cut the gross heating load to an average of 79 kWh/m²a. Passive and internal gains and heat recovery reduce the load that must be met by the heating plant to an average of 20 kWh/m²a. Because some of the houses use a heat pump the average purchased energy for space heating is only 16 kWh/m²a. Figure 3 shows how the heating load is covered.

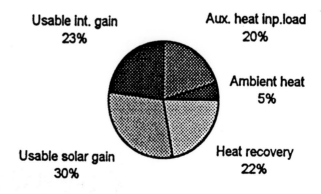

Usable int. gain
23%

Aux. heat inp.load
20%

Ambient heat
5%

Usable solar gain
30%

Heat recovery
22%

Fig. 3: How the heating loads of the IEA houses are met

Minimisation of losses

The calculated heat losses by conduction and infiltration through the building envelope of the houses vary by a factor of nearly three but all values are very low. Figure 4 shows the losses in watts per degree of temperature difference between the inside and outside air. To allow the houses to be compared, the values are per m² of heated floor area.

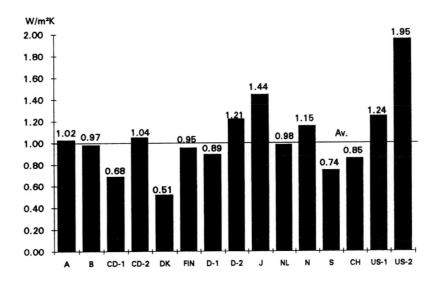

Fig. 4: Heat losses by conduction and air leakage

These very low loss rates result from highly insulated construction. The average U-values of windows is 1.21 W/m²K including the frame and glass. The walls and roofs average 0.18 and 0.13 W/m²K (watts of heat lost per m² of component area and degree of inside-outside temperature difference). As seen in figures 5–7, the houses in colder climates are better insulated than those in milder maritime climates such as Belgium, the Netherlands and Japan.

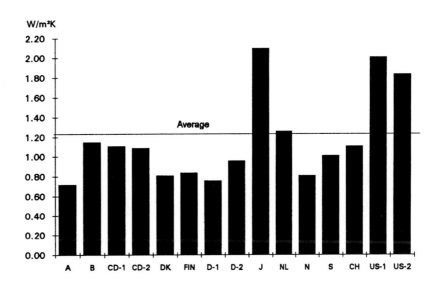

Fig. 5: Window insulation values

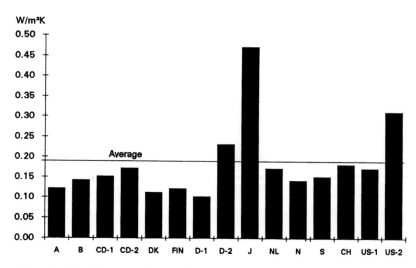

Fig. 6: Wall insulation values

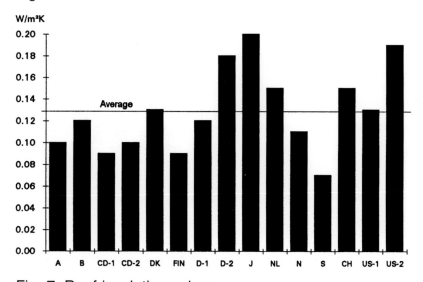

Fig. 7: Roof insulation values

Use of passive solar gains

Passive solar gains contribute on average one-third of the heat the houses require. This exceeds the contribution of the auxiliary heating plant. Three strategies are used to make use of passive solar gains:

1) Sunspaces are included in five of the houses, the largest being the atrium of the Dutch apartment project. This project uses the atrium to preheat ventilation air.

2) Michelle Trombe walls are built in the US-1 house. Masonry below the windows is directly warmed by sunlight. The heat loss to the outside is hindered by glass in front of the walls. The heat that accumulates in the masonry during the day radiates into the rooms later in the night.

3) Direct solar gains through windows are an important aspect of all the houses. Because of the very low U-values

of the windows, solar gains exceed heat losses over the heating season even for east- and west-facing windows in lower latitudes. To increase the usefulness of solar gains, house interiors should include massive materials to store some of the heat. The massiveness of the IEA houses varies from the foam insulation panel construction of the American house to the masonry and concrete construction of the German houses. The Swiss house (CH) is a combination of light exterior walls to maximise insulation and massive interior walls and floors to absorb solar gains. Figure 8 shows the range in the usable passive solar contributions. The peak values for the houses in the Southwest US are not surprising, but the high solar contribution in the German D-1 house is impressive in a much less sunny climate.

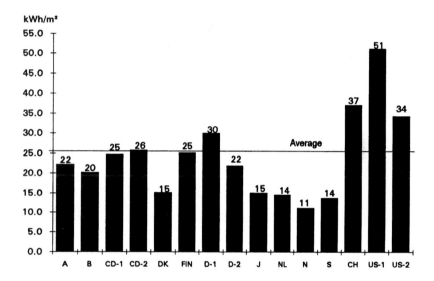

Fig. 8: Usable passive solar gains of the IEA houses

Active solar use

Eight of the houses include active solar water heating systems. The German D-2 house has the largest system, comprising of 40 m² of liquid collectors and very large water storage tanks (19 m³). The excess collection capacity of late summer and autumn can thereby be stored and should meet 100 percent of the winter heating demand.

Heat recovery

On average heat recovery meets 22 percent of the space heating requirement of the houses. All but the Swiss house have a heat recovery system. Heat recovery can be done by a heat exchanger which allows cold intake fresh air to be warmed by a partition separating it from the warm exhaust air. In the Swedish house air channels in the concrete floor serve this function.

Heat pumps

Heat can be extracted from the ground, exhaust air or waste water by a heat pump. This heat can be used for space heating and/or domestic water heating as done in the US, Canadian, Finnish, Japanese and Norwegian houses.

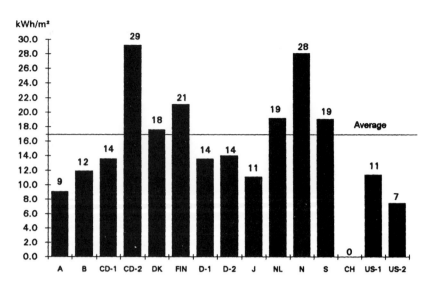

Fig. 9: Heat recovery in the IEA houses

Efficient back-up heating

After the heat input from solar gains, heat recovery, and lights and appliances, very little heat is needed to maintain 20°C room temperature. The average remaining heat demand is 20 kWh/m². Figure 10 illustrates the range in net auxiliary heating load that must be met by the auxiliary heating, and figure 11 shows how much energy the auxiliary heating plant needs to meet this load.

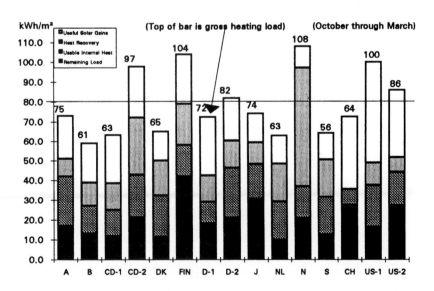

Fig. 10: Back-up heat demand of the IEA houses

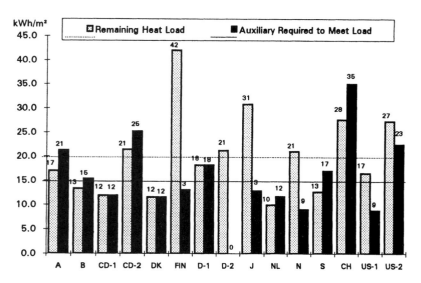

Fig. 11: Heating loads and auxiliary heating energy

The very high remaining heating load and auxiliary heating energy of the Swiss house result from the designer and homeowners not wanting to include mechanical ventilation with heat recovery. Were this feature present, the house's energy performance would fall within the average values of all the houses.

Water heating

The annual heating load for domestic water heating (DHW) is greater than the net space heating load in these houses (23 vs. 20 kWh/m^2). Further, the demand for hot water occurs over the whole year so that the long sunshine hours in summer can be put to full use. Considering these two facts, it is not surprising that nine of the houses incorporate active solar systems to heat domestic hot water. In two cases, the heat is provided by heat recovery.

Electricity consumption

Although the designers focused on the house construction, minimising electrical consumption is also of obvious importance. An annual average value for the household electricity consumption (excluding fans and pumps) was 2500 kWh per year (17 kWh/m^2). To reduce electricity demand, the designers selected the highest efficiency appliances on the market, with particular attention to the refrigerator, range and lighting. The Norwegian, Finnish and Japanese houses are able to cover the household electrical demand with roof-mounted photovoltaic systems. (Annual balance of electricity supplied to and taken from the utility grid).

13

Conclusions

The totals

The goal set for these IEA houses was to achieve very low total purchased energy requirement without compromising comfort. The energy requirements of all the houses is half or less of the levels common for new construction in the respective countries. Space heating remains the largest single energy use (36 percent). Electricity consumption is in second place but if the generation and transmission losses of thermal power stations were considered it would take first place. The energy needed for the fans and pumps of the mechanical ventilation and heat recovery systems is not insignificant (10 percent). Finally, water heating represents a major energy consumption, arguing for solar water heating in very low energy houses.

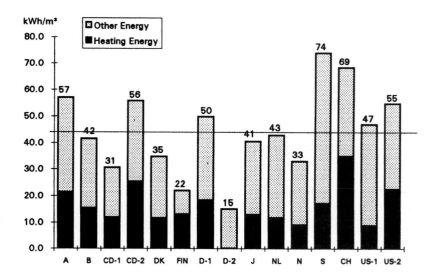

Fig. 12: Total energy demand

These houses are built from readily available materials and components, and the construction is readily done by the trades. The prices of these houses lie within the range between houses with standard builder accessories and those with custom accessories. Accordingly, the energy levels of these IEA houses provide viable bases of comparison for houses being marketed today.

Outlook

Passive solar design, better insulation, tighter construction and ventilation heat recovery will reduce space heating demand so far that electricity and water heating will be the next targets for savings. In the Japanese house electricity use exceeds the total of all other energy demands. Future decreases in the energy consumption in houses will involve conservation and solar systems that serve multiple functions, such as, space heating, water heating and electrical needs.

Austrian Row House in Lustenau

Design team:

Roland Gnaiger, Architect
Kirchstrasse 36
A-6900 Bregenz

Helmut Krapmeier
Energiesparverein Vorarlberg
Institut für Sinnvollen Energie-
einsatz
Postfach 51
A-6851 Dornbirn

I + R Schertler GesmbH
A-6923 Lauterach

Energy features:

Super-insulation
Super-glazing
Direct gain
Solar domestic hot water
Heat recovery
Ground air preheating

Predicted energy demand:
(kWh/m²a)

Space heating	21
Space cooling	0
Water heating	18
Lights & appliances	12
Fans & pumps	6
Total	57

Heated floor area: 85.7 m²

Latitude: 47°N

Solar Radiation (Glob. Vert. S)
Average Monthly Temperature

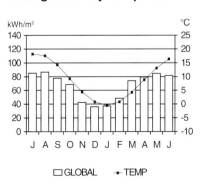

GLOBAL — TEMP

The direct solar gain housing development located in Lustenau (near Lake Constance) must cope with extended periods of fog which can last up to 10 days. This low income housing involves a high density development with small terraced houses, rather than an apartment block.

A mixed form of construction is used. Roofs and external walls consist of well-insulated, lightweight, prefabricated wooden panels between reinforced concrete, sound-insulating party walls. The first floor is solid wood topped by floated concrete.

All main rooms face south looking into the garden in order to maximise both passive solar gains and a sense of neighbourhood. Of the total window area 83 percent is on the south facade which has 26 percent of the total wall area as glass. During the heating season mechanical ventilation is provided. Inlet air is preheated first by a ground heat exchanger and then by a cross-flow heat exchanger. A central wood chip-fired plant, mainly for domestic water heating, provides a backup to the passive solar heating.

MONTHLY ENERGY BALANCES

■ AUX ▨ INT ▥ HEAT REC ▧ SOL (PAS)

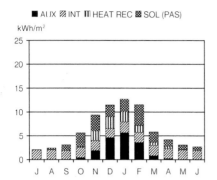

Space Heating

■ AUX ▦ SOL (ACT)

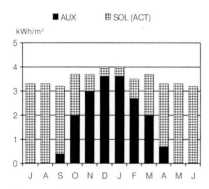

Domestic Hot Water

■ Appliances ▨ Fans & Pumps

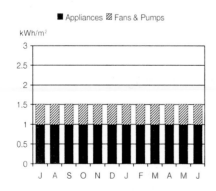

Electricity

U-Values: (W/m²K)

Roof	0.10
Wall	0.12
Window	0.71
Floor	0.19

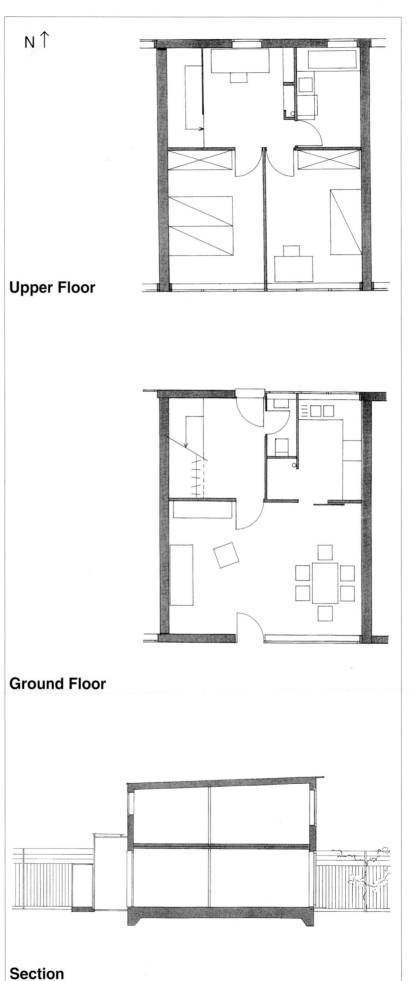

N ↑

Upper Floor

Ground Floor

Section

Belgian Row House in Louvain-la-Neuve

Design team:

SPRL Atelier d'Architecture
Ph. Jaspard
Rue Richier, 45
B-5500 Dinant

Architecture et Climat
Université Catholique de
Place du Levant 1
B-1348 Louvain-la-Neuve

Belgian Building Research
Institute (WTCB/CSTC)
Rue de la Violette 21–23
B-1000 Brussels

Energy features:

Super-insulation
Direct gain
Heat recovery system
Home automation
Solar control

Predicted energy demand:
(kWh/m²a)

Space heating	15
Space cooling	0
Water heating	18
Lights & appliances	7
Fans & pumps	1
Total	41

Heated floor area: 180 m²

Latitude: 51°N

Solar Radiation (Glob. Vert. S)
Average Monthly Temperature

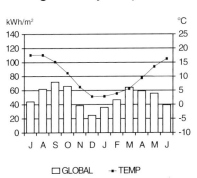

In this two-storey row house, called "Pleiade", built in the new city of Louvain-la-Neuve, special attention is given to the integration of the bioclimatic architectural concepts, the achievement of good thermal comfort in winter and in summer, and good indoor air quality. Accordingly, the design of the envelope was important. Daylighting the central part of this 10 m deep row house was another objective. A balanced ventilation system with heat recovery supplies fresh air to the bedrooms and recirculated air to the living rooms. During hot periods shading of the south-facing glazing and night-time natural ventilation reduces or eliminates overheating.

Two heating systems, a gas air heating system with a four-zone control system, and an electrical heating system were installed for experimental purposes. A home control system ensures optimal energy use and thermal comfort. The system is also used for other functions, such as home security.

MONTHLY ENERGY BALANCES

■ AUX ▨ INT ▥ HEAT REC ▩ SOL (PAS)

kWh/m²

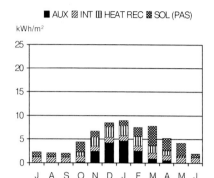

Space Heating

■ AUX

kWh/m²

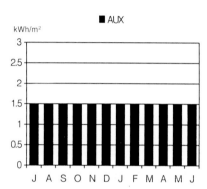

Domestic Hot Water

■ Appliances ▨ Fans & Pumps

kWh/m²

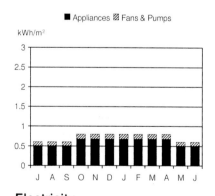

Electricity

U-Values: (W/m²K)

Roof	0.12
Wall	0.14
Window	1.14
Window including frame	1.49
House to basement	0.19
Floor	0.19

N↑

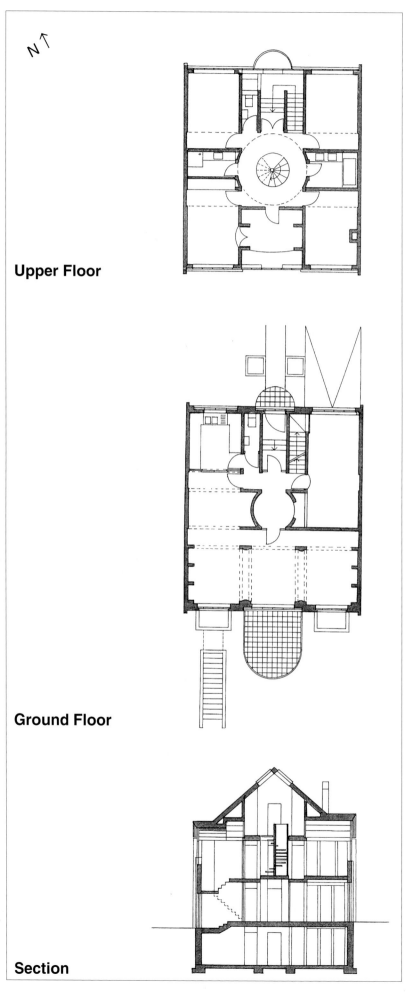

Upper Floor

Ground Floor

Section

Canadian Single-family House in Brampton

Design team:

Elizabeth White, Architect
Marsh Hill Farm R.R.4
CDN-Stirling
Ontario K0K 3E0

Greg Allen (Integrated
Mechanical System)
Allen Associates
400 Mount Pleasant Road
Suite 5
CDN-Toronto
Ontario M4S 2L6

Energy features:

Super-insulation
Super-glazing
Integrated mechanical system
Sunspace preheat. of ventil.
air
Energy-efficient appliances
and lighting

Energy demand:
(kWh/m²a)

Space heating	12
Space cooling	1
Water heating	5
Lights & appliances	10
Fans & pumps	3
Total	31

Heated floor area: 408 m²

Latitude: 44°N

Solar Radiation (Glob. Vert. S)
Average Monthly Temperature

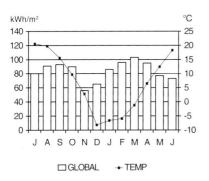

The building shell of the Brampton Advanced House is light-weight, well insulated and airtight. Walls are wood frame with 240 mm of wet-blown cellulose insulation and 25 mm of rigid fibreglass; the ceilings are insulated with 450 mm of blown cellulose. The concrete basement walls are insulated with 175 mm of blown cellulose and 50 mm of rigid insulation under the concrete floor. The wood-framed windows are triple-glazed with two low-e coatings, argon gas filled and have insulated spacers. A two-storey sunspace preheats ventilation air through ducts in the sunspace wall and floor slab.

An integrated mechanical system provides heating, cooling, ventilation and water heating. The system has a hot water tank and an ice storage tank. Heat, recovered from exhaust air, grey water and excess passive solar gains is fed to the ice storage tank. A heat pump upgrades this heat for the hot water tank which supplies space and water heating. Fluid from the ice tank is circulated through a fan coil to provide summer air conditioning.

Energy-efficient appliances and lighting use only half the energy conventionally used. The full-size refrigerator consumes less than 30 kWh per month. All lighting is fluorescent and halogen; no incandescent lighting is used.

MONTHLY ENERGY BALANCES

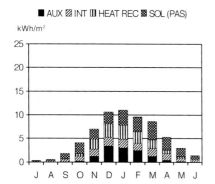

■ AUX ▨ INT ⬚ HEAT REC ▨ SOL (PAS)

kWh/m²

Space Heating

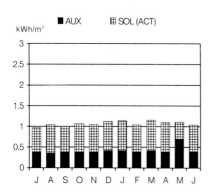

■ AUX ⊞ SOL (ACT)

kWh/m²

Domestic Hot Water

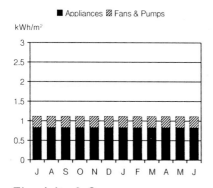

■ Appliances ▨ Fans & Pumps

kWh/m²

Elecricity & Gas

U-Values: (W/m²K)

Roof	0.09
Wall	
to amb.	0.15
to earth	0.15
to sunspace	0.48
Window	1.10
Sunspace	1.00
Floor/earth	0.63

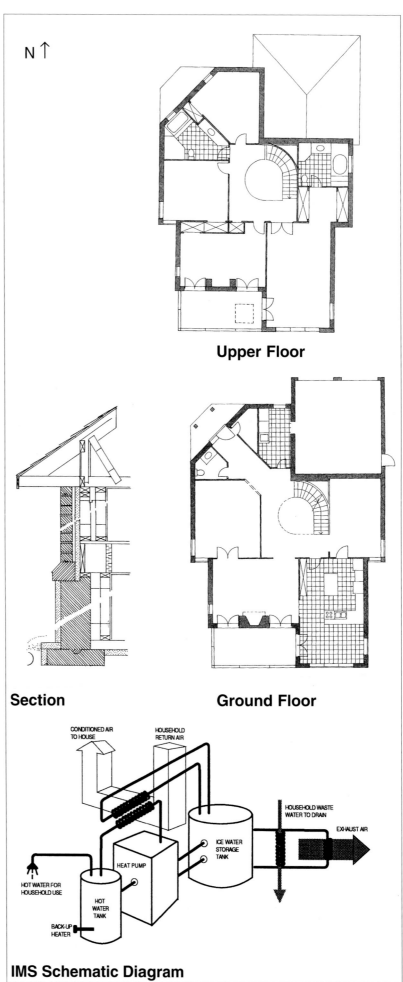

N ↑

Upper Floor

Section

Ground Floor

CONDITIONED AIR TO HOUSE

HOUSEHOLD RETURN AIR

HOUSEHOLD WASTE WATER TO DRAIN

EXHAUST AIR

ICE WATER STORAGE TANK

HEAT PUMP

HOT WATER FOR HOUSEHOLD USE

HOT WATER TANK

BACK-UP HEATER

IMS Schematic Diagram

Canadian Single-family House in Waterloo

Design team:

Enermodal Engineering
368 Phillip Street, Unit 2
CDN-Waterloo
Ontario N2L 5J1

Richard Reichard
Snider Reichard March
Architects
145 Columbia Street West
CDN-Waterloo
Ontario N2L 3L2

Energy features:

PV-pumped Solar DHW
Combined furnace and heat recovery system
Super-glazing
Cistern/ground cooling
Recycled materials
Energy-efficient appliances and lighting

Predicted energy demand:
(kWh/m²a)

Space heating	25
Space cooling	1
Water heating	5
Lights & appliances	17
Fans & pumps	8
Total	56

Heated floor area: 208 m²

Latitude: 44°N

Solar Radiation (Glob. Vert. S)
Average Monthly Temperature

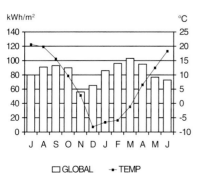

The Waterloo Region Green House has a well-insulated and airtight shell. Above-grade walls are framed with wooden I-beams and below-grade walls use a precast concrete system requiring only half the concrete of typical foundations. The triple-glazed windows feature insulated fibreglass frames. The house finishes include reused wood flooring, refurbished bathroom fixtures, recycled gypsum board, carpets made from plastic pop bottles, and formaldehyde- and VOC-free paints, glues and fabrics.

A prototype combination natural gas furnace and air-to-air heat exchanger of small rocks provides heating and mechanical ventilation. A solar domestic hot water system, with the circulation pump powered by a PV panel, meets most of the hot water needs. Water conservation features include a cistern, 3 l flush toilets, and low water use appliances. Energy-efficient appliances include a direct-vent natural gas stove and a non-CFC refrigerator.

Operable clerestory windows and a thermochromatic film (that turns from clear to translucent above 26°C) reduce the need for cooling. Cistern water is circulated through plastic tubing buried below the basement and through a coil in the air-handling system to provide summer cooling.

MONTHLY ENERGY BALANCES

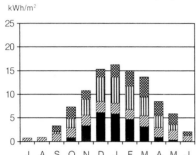

■ AUX ▨ INT ▥ HEAT REC ▩ SOL (PAS)

kWh/m²

Space Heating

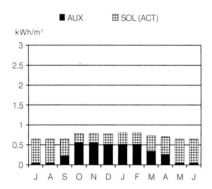

■ AUX ⊞ SOL (ACT)

kWh/m²

Domestic Hot Water

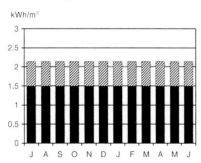

■ Appliances ▨ Fans & Pumps

kWh/m²

Electricity & Gas

U-Values: (W/m²K)

Roof	0.10
Wall	
to amb.	0.17
to earth	0.22
Window	1.08
Centre Glazing	0.97
Floor	0.80

N ↑

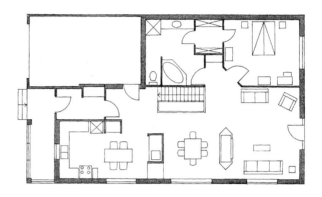

Upper Floor

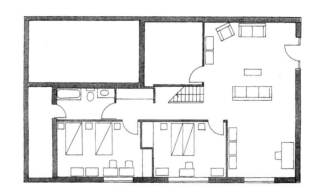

Ground Floor

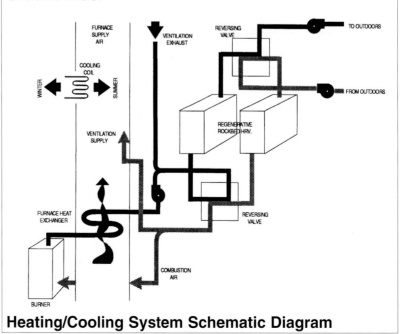

Heating/Cooling System Schematic Diagram

Danish Row Houses at Kolding

Design team:

Boje Lundgaard & Lene
Tranberg , Architects
St. Kannikestraede 6
DK-1169 Copenhagen K

Thermal Insulation Laboratory
Technical University (DTU)
Building 118
DK-2800 Lyngby

Energy features:

Super-insulation
Super-windows
Solar domestic hot water
Fast -responding heating
Heat recovery
Energy-efficient appliances
Energy-efficient lighting

Predicted energy demand:

(kWh/m²a)	Type 1	Type 2
Space heating	14	12
Space cooling	0	0
Water heating	5	5
Lights & appliances	12	13
Fans & pumps	5	5
Total	36	35

Heated floor area:
Type 1: 105 m²
Type 2:   98 m²

Latitude: 55° N

**Solar Radiation (Glob. Vert. S)
Average Monthly Temperature**

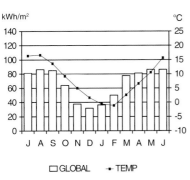

These row houses, planned for south-east Jutland, about 200 km west of Copenhagen, have been designed as solutions for different site orientations. Type 1 is a solution for house rows running north–south, Type 2 is for rows running east–west (reported in the statistics in the overview chapter). Interestingly, in both concepts it is possible to reach extremely low heating energy demands using nearly the same active and passive solar features.

In both house types the same envelope construction, heating, ventilation, and active solar systems are used. The opaque envelope is insulated with 300 mm mineral wool. Windows have a low-U glazing and insulated wooden frames. Aerogel is used in skylights.

The heating system was designed to respond quickly to solar and internal gains. Fan-driven convectors are controlled by room thermostats. The small, highly efficient, low water content gas burner has a 400 l domestic hot water tank as a buffer. A mechanical system with balanced ventilation and a counterflow heat exchanger achieves 80 percent heat recovery. An active solar system covers nearly 80 percent of the domestic hot water demand. Water- and electricity-saving devices and low energy appliances were selected.

MONTHLY ENERGY BALANCES

■ AUX ▨ INT Ⅲ HEAT REC ▧ SOL (PAS)

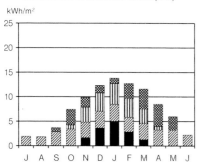

Space Heating   Type 1

■ AUX ▨ INT Ⅲ HEAT REC ▧ SOL (PAS)

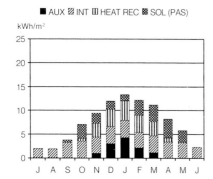

Space Heating   Type 2

■ AUX   ▦ SOL (ACT)

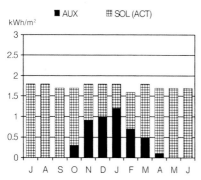

Domestic Hot Water

■ Appliance ▨ Fans & Pumps

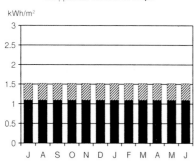

Electricity

U-values: (W/m²K)

Roof	0.13
Windows	0.80
Skylight	0.50
Floor	0.10
Wall	0.11

North-south row   East-west row

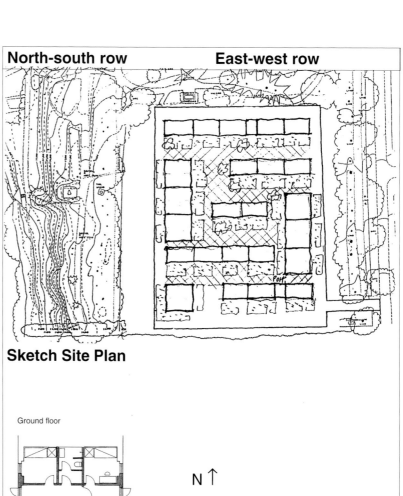

Sketch Site Plan

Ground floor

N ↑

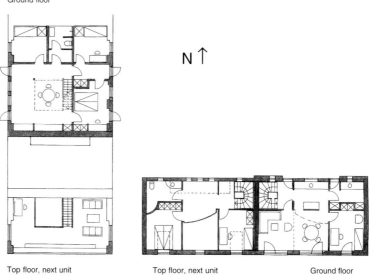

Top floor, next unit   Top floor, next unit   Ground floor

Type 1 (N/S)   Type 2 (E/W)

Floor Plans

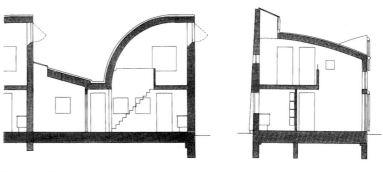

Type 1 (N/S)   Type 2 (E/W)

Cross Sections

Finnish Single-family House in Pietarsaari

Design team:

Ilpo Kouhia
VTT Building Technology
P.B. 18011
FIN-02044 VTT

Andreas Walterman
ALW-ARK Oy
Lastenkodinkatu 5
FIN-65100 Vaasa

Neste Advanced Power
Systems
Rälssitie 7
FIN-01510 Vantaa

Energy features:

Super-insulation
Super-windows
Ground heat pump
Solar hot water
PV systems

Predicted energy demand:
(kWh/m²a)

Space heating	13
Space cooling	0
Water heating	12
Lights & appliances	12
Fans & pumps	4
PV electricity	-14
Total	27

Heated floor area: 166 m²

Latitude: 62°N

Solar Radiation (Glob. Vert. S)
Average Monthly Temperature

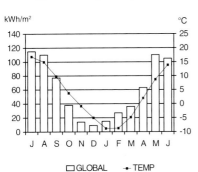

☐ GLOBAL   ▪ TEMP

This two-storey house, located in Pietarsaari (Central Finland) is built in lightweight wooden construction with large prefabricated panels. The building envelope is highly insulated and is more airtight than conventional construction. Energy sources include a 2 kW photovoltaic system, solar thermal collectors and a ground-coupled heat pump. The small sunspace on the south facade provides only minimal energy savings.

Space heating is provided by a low-temperature floor heating system, incorporating a 3000 l water tank for heat storage. In the winter the tank is mainly charged by the heat pump (COP 3.3); at other times it is charged by the solar collectors. The ground piping loops of the heat pump can be used for cooling the tank if it overheats in summer. The floor heating system can be used for cooling as well.

Domestic hot water is heated in the upper section of the tank. If necessary an auxiliary heater in a separate 60 l tank is available. The mechanical ventilation system of the building includes two heat recovery units, one for the kitchen hood and one for the bathrooms, with expected efficiencies of 60 and 80 percent respectively.

MONTHLY ENERGY BALANCES

INT ☐ HEAT REC ☐ SOL (PAS) ☐
SOL (ACT) ☐ HEAT PUMP ☐

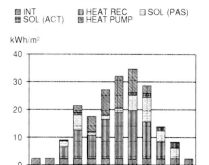

Space Heating

SOL (ACT) ■ HEAT PUMP ☐

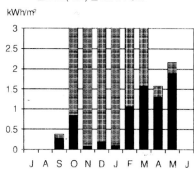

Domestic Hot Water

Appliances ☐ Heat Pump ☐

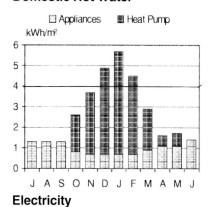

Electricity

AUX ■ PV SOLAR ☐

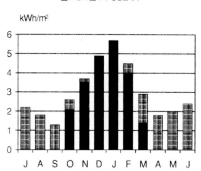

Total Electricity

U-Values: (W/m²K)

Roof	0.09
Wall	0.12
Window	0.70
Floor	0.11
Sunspace	1.80

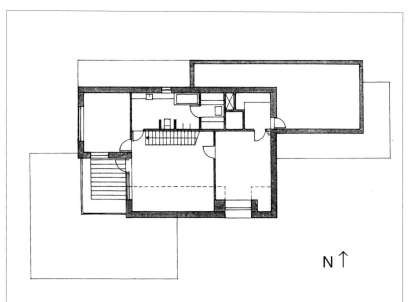

N ↑

Upper Floor

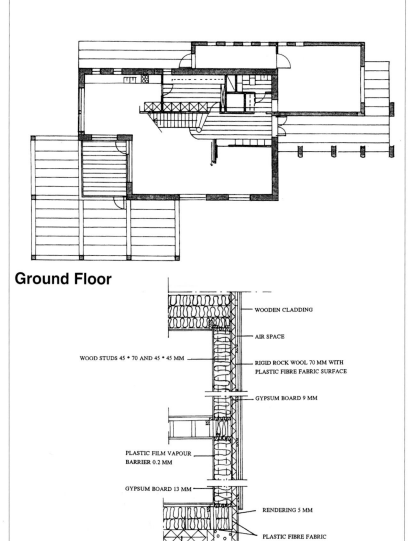

Ground Floor

WOODEN CLADDING

AIR SPACE

WOOD STUDS 45 * 70 AND 45 * 45 MM

RIGID ROCK WOOL 70 MM WITH PLASTIC FIBRE FABRIC SURFACE

GYPSUM BOARD 9 MM

PLASTIC FILM VAPOUR BARRIER 0.2 MM

GYPSUM BOARD 13 MM

RENDERING 5 MM

PLASTIC FIBRE FABRIC

RIGID ROCK WOOL 100 + 100 MM WITH PLASTIC FIBRE FABRIC SURFACE

RIGID ROCK WOOL 100 MM

Section

German Duplex in Rottweil

Design team:

Institut für Bau-, Umwelt und Solarforschung
Caspar Theyss Strasse 14A
D-14193 Berlin

Fraunhofer Institut für Bauphysik
Postfach 800469
D-70569 Stuttgart

Energy features:

Super-insulation
Super-glazing
Ground air preheating
Heat recovery

Predicted energy demand:
(kWh/m²a)

Space heating	18
Space cooling	0
Water heating	15
Lights & appliances	11
Fans & pumps	3
Total	50

Heated floor area: 175 m²

Latitude: 48°N

Solar Radiation (Glob. Vert. S)
Average Monthly Temperature

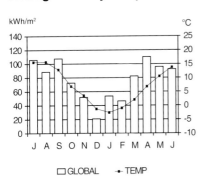

The "ultra low energy house" is a duplex located in Rottweil (50 km south of Stuttgart). The west-facing half of the house was constructed as an advanced low energy house, while the east half was constructed as a low energy house (annual space heating: 49 kWh/m²) and provides a useful comparison.

The building opens like a fan with maximum southern exposure (containing c. 90 percent of the glazing) and minimum northern exposure. A multi-storied sunspace is embedded in the south side of each house. The roof has 340 mm of insulation and walls have 400 mm. The window glazing is filled with xenon and achieves a U-value of 0.5 W/m²K.

Waste hot water from a local generator power plant provides space and domestic water heating via a heat exchanger. Heat is delivered to the rooms by radiators. The efficiency of the heat recovery exceeds 75 percent. When air temperatures are less than 8°C, the air is preheated in subterranean pipes before passing through a heat exchanger to reclaim heat from exhaust air. For temperatures between 8°C and 15°C, only the heat exchanger is used. Between 15°C and 25°C, both heat exchanger and the ground system are by-passed. When air temperatures exceed 25°C, the ground system is used for cooling.

MONTHLY ENERGY BALANCES

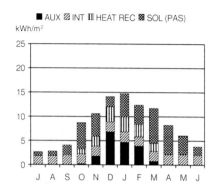

■ AUX ▨ INT Ⅲ HEAT REC ▩ SOL (PAS)

kWh/m²

Space Heating

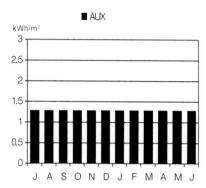

■ AUX

kWh/m²

Domestic Hot Water

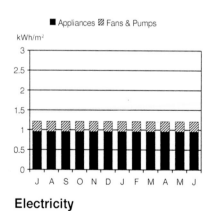

■ Appliances ▨ Fans & Pumps

kWh/m²

Electricity

U-Values: (W/m²K)

Roof	0.12
Wall amb.	0.10
Wall cellar	
to earth	0.30
to living	0.20
Window	0.75
Floor	
living to earth	0.15
cellar to earth	0.32
cons. to earth	0.17

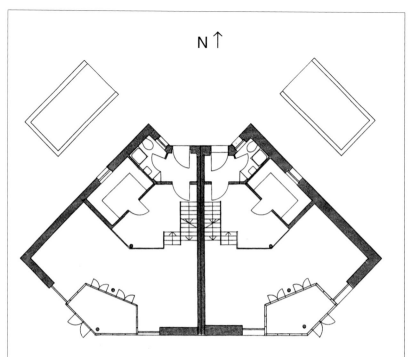

N ↑

Ground Floor

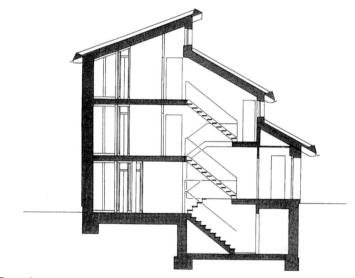

Section

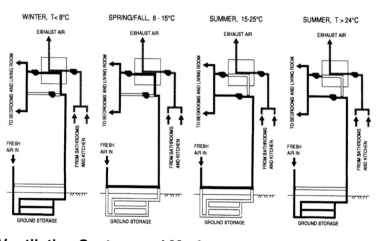

Ventilation System and Mode

German Row House in Berlin

Design team:

Institut für Bau-, Umwelt-
und Solarforschung
Caspar Theyss Strasse 14A
D-14193 Berlin

Fraunhofer Institut für Bau-
physik
Postfach 800469
D-70569 Stuttgart

Energy features:

Solar hot water system
Seasonal heat storage
Super-glazing
Heat recovery

Predicted energy demand:
(kWh/m² a)

Space heating	0
Space cooling	0
Water heating	0
Lights & appliances	12
Fans & pumps	3
Total	15

Heated floor area: 170 m²

Latitude: 52°N

Solar Radiation (Glob. Vert. S)
Average Monthly Temperature

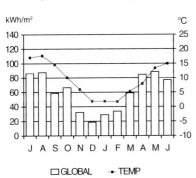

□ GLOBAL -•- TEMP

This German Zero Energy House is located in Spandau, Berlin. Its situation, at the corner of two rows of houses, affords a large south facade and a minimal north facade. The three storey south-facing sloped facade has window bands to receive direct solar gains and a seasonal storage system incorporated into the building structure.

The building is well insulated with the large glazing area on the southern side having windows with a U-value of 0.95 W/m²K. The roof has 180 mm mineral wool, walls have 160 mm and the floor above the basement has 120 mm. The ventilation system includes heat recovery and has two speeds: normal operation and forced ventilation of kitchen and baths for periods of high humidity production.

Two water tanks, of 11 m³ and 8 m³, placed in the middle of the house and in the basement, store the heat from solar collectors. Heat losses from the tanks (insulated with 300 mm of mineral wool), can be useful for space heating. Space heating is by low volume highly efficient radiators. The fluid within the collectors circulates by natural convection, but a small pump is used to overcome the resistance of the heat exchanger coils.

MONTHLY ENERGY BALANCES

◪ INT ▥ HEAT REC ▨ SOL (PAS) ⊞ SOL (ACT)

kWh/m²

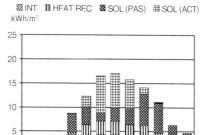

Space Heating

⊞ SOL (ACT)

kWh/m²

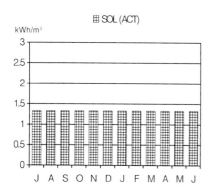

Domestic Hot Water

■ Appliances ▨ Fans & Pumps

kWh/m²

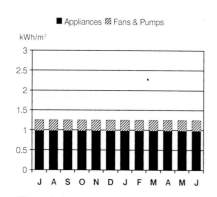

Electricity

U-Value: (W/m²K)

Roof	0.19
Window	0.95
Floor	0.31
Wall amb.	0.20
Wall coll.	0.24
Wall cellar.	
to earth	0.20
to amb.	0.20

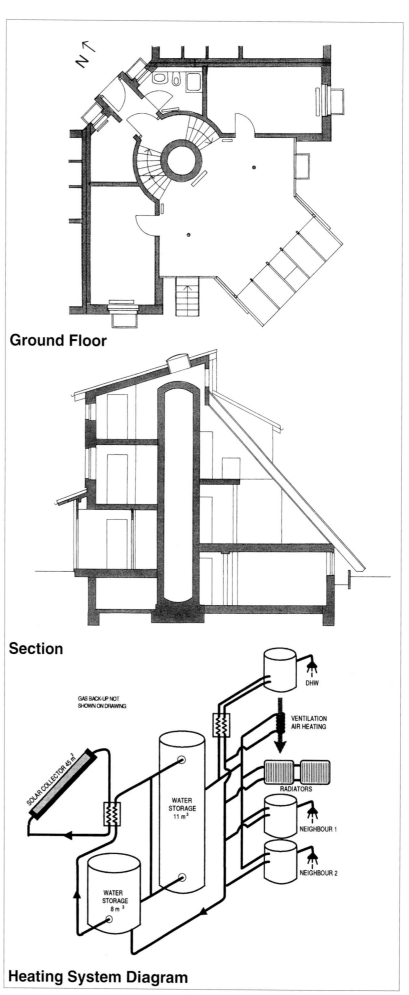

Ground Floor

Section

GAS BACK-UP NOT
SHOWN ON DRAWING

SOLAR COLLECTOR 45 m²

WATER STORAGE 11 m³

WATER STORAGE 8 m³

DHW

VENTILATION AIR HEATING

RADIATORS

NEIGHBOUR 1

NEIGHBOUR 2

Heating System Diagram

Japanese Duplex in Iwaki

Design team:

Sekisui House Ltd.
Tokyo Planning Office
11-3 Nishi-Shinjuku-1
Shinjuku-ku
J-Tokyo 160

Ken-ichi Kimura
Dep. of Architecture
Waseda University
4–1 Okubo 3, Shinjuku
J-169 Tokyo

Mitsuhiro Udagawa
Department of Architecture
Kogakuin University
24–2 Nishi-Shinjuku-1
Shinjuku-ku
J-Tokyo 160

Predicted energy features:

PV array system
Heat pump system
PCM panels

Energy demand:
(kWh/m^2 a)

Space heating	13
Space cooling	7
Water heating	11
Lights & appliances	38
Fans & pumps	6
PV electricity	-37
Total	38

Heated floor area: 150 m^2

Latitude: 37 °N

**Solar Radiation (Glob. Vert. S)
Average Monthly Temperature**

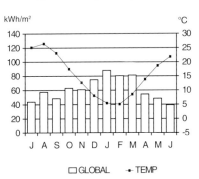

The "Dream House" is situated in Iwaki city close to the Pacific Ocean 180 km north of Tokyo. Because it was planned for a family of two generations it is divided into two units. The south unit is for grandparents, and the north unit is for a couple with two children. The north unit, with its various innovative systems, is reported here.

The walls are insulated with 50 mm urethane foam, the floor with 100 mm polystyrene foam and the roof with 220 mm mineral wool. All windows have low-e coated argon gas filled double glazing and plastic frames.

A hybrid PV collector (4.2 kW peak) is installed on the roof. Air warmed by the PV modules is a heat source of a heat pump for supplying space heating. Most of the heating and cooling loads should be covered by phase change material (PCM) panels embedded in the ceiling and floor. An air-conditioning system containing a heat recovery ventilation unit backs up the PCM panels. Because the PCM panels can only remove sensible heat, a dehumidifying unit impregnated with lithium chloride solution is also used. A 6 m^2 flat plate solar collector array with a 300 l tank will cover one-half of the domestic hot water demand. The total purchased energy of the north unit is estimated to be about 40 percent of that of an average single-family house in Japan.

31

MONTHLY ENERGY BALANCES

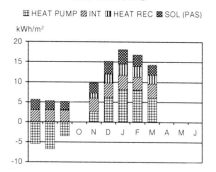

⊞ HEAT PUMP ▨ INT ▥ HEAT REC ▩ SOL (PAS)

Space Heating

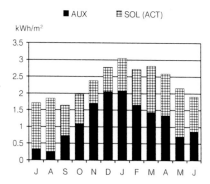

■ AUX ⊞ SOL (ACT)

Domestic Hot Water

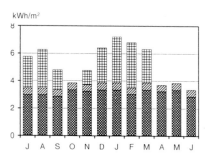

▩ Appliances ▨ Fans & Pumps ⊞ Heat Pump

Electricity

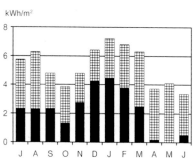

■ AUX ⊞ PV SOLAR

Total Electricity

U-Values: (W/m²K)

Wall	0.47
Window	2.09
Ceiling	0.20
Floor	0.32

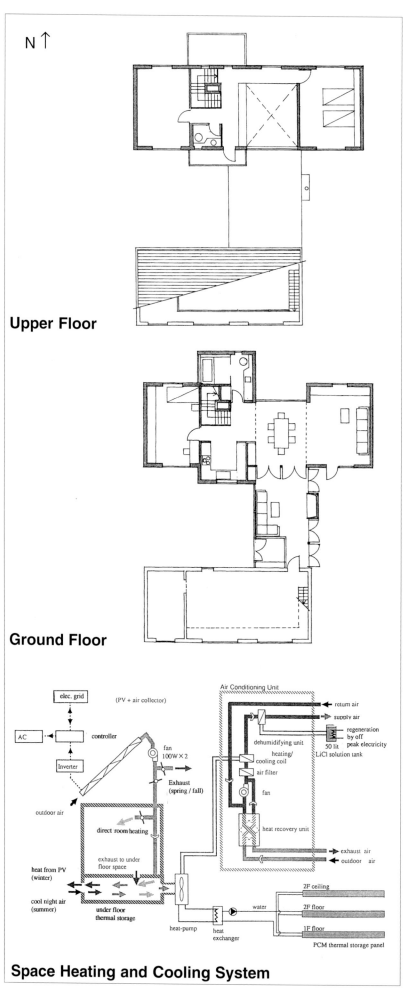

N ↑

Upper Floor

Ground Floor

Space Heating and Cooling System

The Netherlands Apartment Building in Amstelveen

Design team:

Atelier Z
Zavrel Architecten BV
St. Jobsweg 30
NL-3024 EJ Rotterdam

Damen Consultants
Postbox 694
NL-6800 AR Arnhem

Delft Technical University
Mechanical Engineering dept.
Mekelweg 2
NL-2628 CD Delft

Energy features:

Super-insulation
Direct gain
Passive cooling
Solar domestic hot water
High efficiency heat recovery
Advanced control system

Predicted energy demand:
(kWh/m²a)

Space heating	12
Space cooling	0
Water heating	11
Lights & appliances	16
Fans & pumps	4
Total	43

Heated floor area: 100 m²

Latitude: 52°N

Solar Radiation (Glob. Vert. S)
Average Monthly Temperature

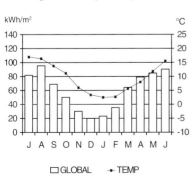

This 42-unit apartment building has been built in a new development area in Amstelveen (near Amsterdam). It consists of two four-storey buildings and a five-storey building connected by a large atrium. In each of the south-facing lower buildings, special care was taken to reduce losses, optimise solar gains and avoid overheating.

The entire thermal envelope is highly insulated. Of special interest is the prefabricated south facade. It incorporates direct gain windows, a passive cooling system, shading devices and auxiliary heating systems.

The ventilation system incorporates an advanced heat recovery unit. Domestic hot water is preheated by solar collectors on the roof of the same building. Backup water and space heating are provided by a high efficiency gas boiler. Finally, a control system ensures that components optimally interact with one another. Advanced control systems in each apartment prevent a room from being heated without heat recovery. A central control system provides coordination and data processing as well as managing the ventilation system.

MONTHLY ENERGY BALANCES

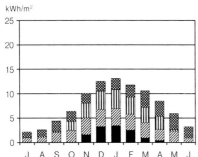

Space Heating

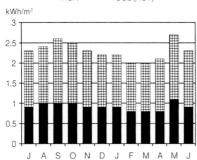

Domestic Hot Water

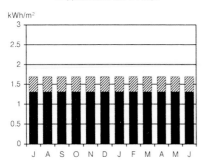

Electricity

U-Values: (W/m²K)

Roof	0.15
Wall	0.17
Window	0.70
Floor	0.15
Atrium	5.70

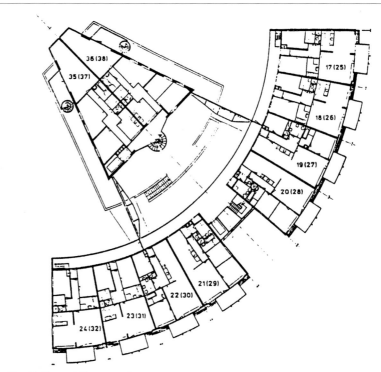

Building Floor Plan

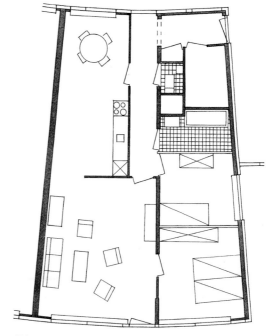

Apartment Floor Plan

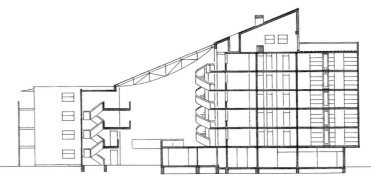

Cross Section

Norwegian Row House at Hamar

Design team:

SINTEF:
Divisions of Architecture and
Building Technology, Applied
Thermodynamics and
Refrigeration Engineering
N-7034 Trondheim

NTH:
Department of Building
Technology, Norwegian
Institute of Technology
N-7034 Trondheim

Energy features:

Super-insulation
Solar assisted ground heat
pump
Heat recovery
PV system
Sunspace

Predicted energy demand:
(kWh/m²a)

Space heating	9
Space cooling	0
Water heating	13
Lights & appliances	20
Fans & pumps	14
PV electricity	-22
Total	34

Heated floor area: 125 m²

Latitude: 61°N

Solar Radiation (Glob. Vert. S)
Average Monthly Temperature

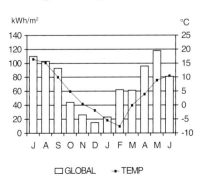

□ GLOBAL   ←TEMP

The IEA dwelling is the centre unit of a three-unit row house at Hamar, near Lillehammer.

The heating demand has been minimised by use of light super-insulated constructions in the envelope, windows with super-glazing and insulated frames, and a south-facing sunspace with a capillary-type transparent insulation in the double-glazed window units. The house features a mechanical ventilation system with balanced ventilation and a rotating wheel heat recovery unit with an efficiency of 90 percent.

All space and domestic hot water heating are supplied by a solar assisted heat pump (COP 2.5) that uses the ground and the small sunspace as heat sources, with a ground loop and an uncovered absorber in the sunspace.

A part of the electricity demand in the house is met by a photovoltaic system that is coupled to the grid. The system is expected to deliver surplus electricity to the utility during sunny periods in the summer months.

The total auxiliary energy needed is about 15 percent of that needed in a similar dwelling built according to the existing building code.

MONTHLY ENERGY BALANCES

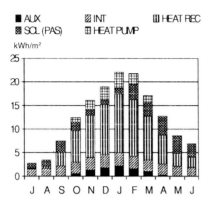

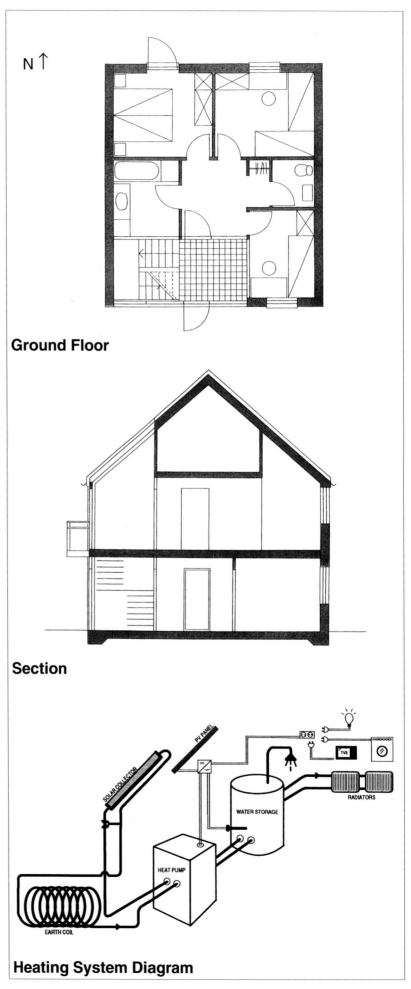

■ AUX   ▨ INT   ⊞ HEAT REC
▨ SOL (PAS)   ⊞ HEAT PUMP

kWh/m²

Space Heating

■ AUX   ⊞ HEAT PUMP

kWh/m²

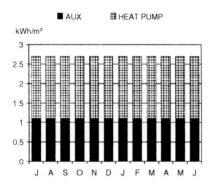

Domestic Hot Water

■ Appliances   ▨ Fans & Pumps

kWh/m²

Electricity

U-values (W/m²K):

Roof	0.11
Wall	
to ambient	0.14
uncond. space	0.60
Window	
to ambient	0.80
to sunspace	0.9–1.5
Sunspace	0.6–0.8
Floor/earth	0.13

N ↑

Ground Floor

Section

Heating System Diagram

Swedish Single-family House in Röskär

Design team:

Gudni Johannesson
Dept. of Building Tech.
KTH
S-10044 Stockholm

Carl Michael Johannesson

Göran Werner

Energy features:

Super-insulation
Super-glazing
Hollow core floor heat
Heat recovery
Solar DHW

Predicted energy demand:
(kWh/m²a)

Space heating	17
Space cooling	0
Water heating	21
Lights & appliances	25
Fans & pumps	11
Total	74

Heated floor area: 50.4 m²

Latitude: 60°N

Solar Radiation (Glob. Vert. S)
Average Monthly Temperature

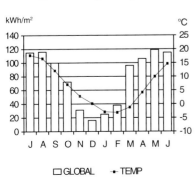

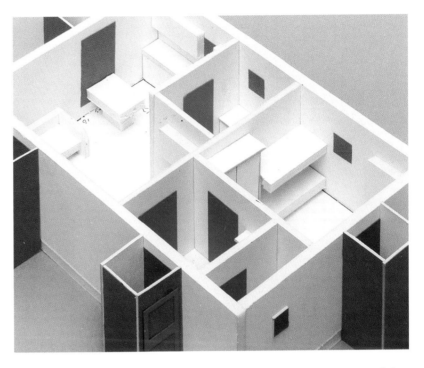

This modified Swedish farmhouse layout from the turn of the last century is based on block house construction. Two features of the house are unique: a hybrid floor heating system and a variable window.

The concrete floor slab incorporates air channels and hot water piping. The floor heating operates with an inlet temperature of 30°C. The ventilation air is preheated in series with the floor piping. During the heating season the domestic hot water is also preheated from 10° to 30°C from the floor, then further heated to 55°C by a bottle gas boiler.

The window system is comprised of a quadratic glazed box mounted outside the normal window. Within the box is a rigid insulating panel with reflecting surfaces. The panel can be rotated by a small 12 V electric motor. When the reflector is angled inwards, it increases the solar gain through the window. In position on the side towards the sun, it acts as a sunshade. In a third position, it acts as a night insulation, decreasing the window U-value to that of a wall.

MONTHLY ENERGY BALANCES

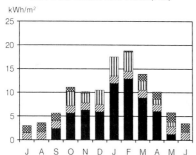

■ AUX ▨ INT ▥ HEAT REC ▧ SOL (PAS)

kWh/m²

Space Heating

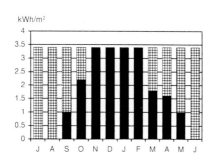

■ AUX ▦ SOL (ACT)

kWh/m²

Domestic Hot Water

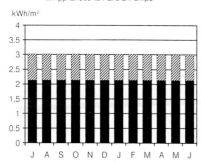

■ Appliances ▨ Fans & Pumps

kWh/m²

Electricity

U-values (W/m·K):

Roof	0.07
Wall	0.15
Window	1.00
Floor/earth	0.15

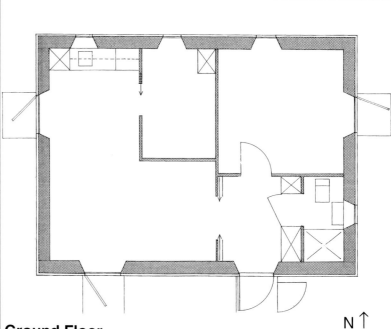

Ground Floor

N ↑

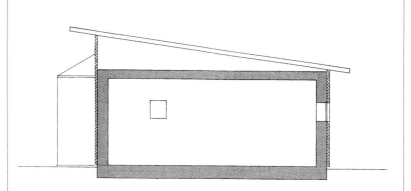

Section

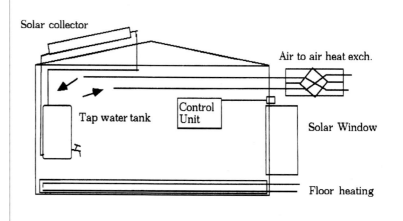

System Diagram

Swiss Duplex in Gelterkinden

Design team:

Ueli Schäfer
Dipl. Architekt BSA/SIA
Zollikonstrasse 20
CH-8122 Binz

Solararchitektur
Energy analyst
ETH Hönggerberg
CH-8093 Zürich

Energy features:

Direct gain
Super insulation
High performance windows

Predicted energy demand:
(kWh/m²a)

Space heating	28
Space cooling	0
Water heating	18
Light & appliances	11
Fans & pumps	0
Total	57

Heated floor area: 182 m²

Latitude: 47°N

Solar Radiation (Glob. Vert. S)
Average Monthly Temperature

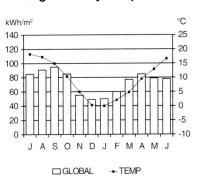

The Chienbergreben Housing Development (located near Basel) is on a south slope, formerly a vineyard. Each semi-detached house has an optimal south exposure. The south facade of every unit has a window band extending from the ground to the roof. The open plans of the ground and upper floors avoid the overheating that often occurs in small, closed rooms with large south-facing windows. Because the house shell is a light wooden construction, interior partitions and floors are massive and partially of stone, to provide the needed thermal mass.

A central woodchip heating plant provides backup heat distributed by a two-pipe hot water system coupled with a 300 l water tank in each house.

The predicted auxiliary space heating demand is low (28 kWh/m²a). The large glass area is responsible for 38 percent of the heating demand. Heating supply air accounts for 35 percent of the heating load but a conscious decision was made not to mechanically ventilate the houses. Solar gains cover 51 percent of the heating load.

MONTHLY ENERGY BALANCES

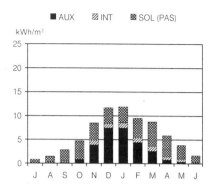

■ AUX   ▨ INT   ▧ SOL (PAS)

kWh/m²

Space Heating

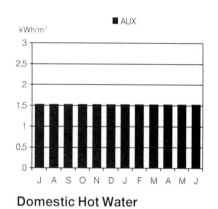

■ AUX

kWh/m²

Domestic Hot Water

■ Appliances

kWh/m²

Electricity

U-Values: (W/m²K)

Roof	0.15
Wall	0.18
Window (centre of glass)	1.1
Floor (unheated basement)	0.34

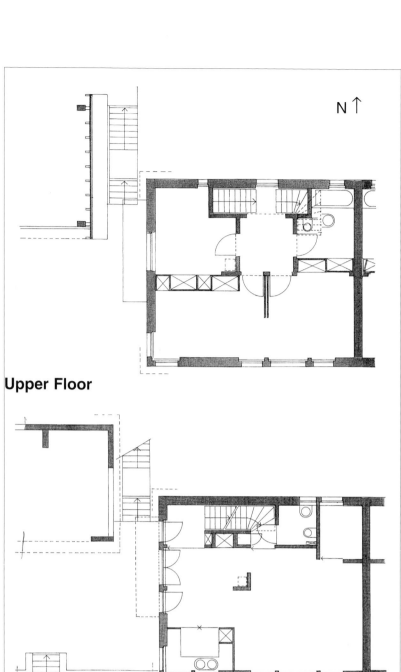

N ↑

Upper Floor

Ground Floor

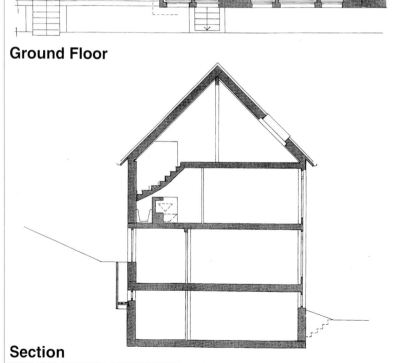

Section

American Single-family House at Grand Canyon

Design team:

U.S. National Park Service

OZ Architecture
1580 Lincoln, Suite 200
USA-80203 Denver CO

J. Douglas Balcomb
National Renewable Energy
Laboratory (NREL)
1617 Cole Blvd
USA-80401 Golden CO

Energy features:

Direct gain
Michelle Trombe walls
Structural insulated panels
Integrated mechanical system
Efficient appliances and lighting

Predicted energy demand:
(kWh/m²a)

Space heating	10
Space cooling	0
Water heating	12
Lights & appliances	19
Fans & pumps	6
Total	47

Heated floor area: 125 m²

Latitude: 36°N

Solar Radiation (Glob. Vert. S)
Average Monthly Temperature

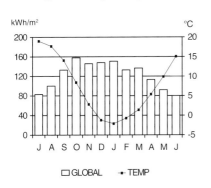

☐ GLOBAL   -•- TEMP

This single-family house near the south rim of the Grand Canyon is at an altitude of 2100 m. The combination of an extremely cold winter climate with abundant sunshine provides an ideal climatic design opportunity.

Passive solar gains are provided by 11.3 m² of south-facing windows. Beneath the windows are 7.5 m² of short, unvented Michelle Trombe walls constructed of 200 mm thick concrete block, an IR-selective coated foil absorber to reduce thermal losses and double glazing of uncoated water-white glass to maximise solar transmission.

Walls and roof are factory-assembled structural panels with a polystyrene foam core (walls 152 mm and roof 254 mm) sandwiched between strandboard stress-skins. The splined and glued panels provides a structure that is four times more rigid than conventional construction, has few thermal bridges and is quite airtight.

An integrated heat pump system with a COP of 3.1 provides controlled ventilation, heat for all the domestic hot water, and 70 percent of the back-up space heat. Electricity consumption is half that of a normal household, thanks to efficient lighting and an energy-efficient refrigerator.

MONTHLY ENERGY BALANCES

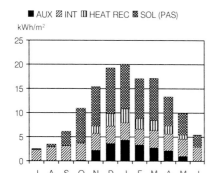

Space Heating

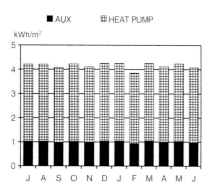

Domestic Hot Water

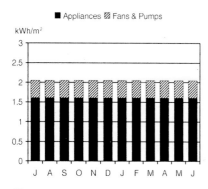

Electricity

U-Values: (W/m²K)

Roof	0.10
Wall	
to amb.	0.17
to earth	0.22
Window	1.08
Floor	0.80

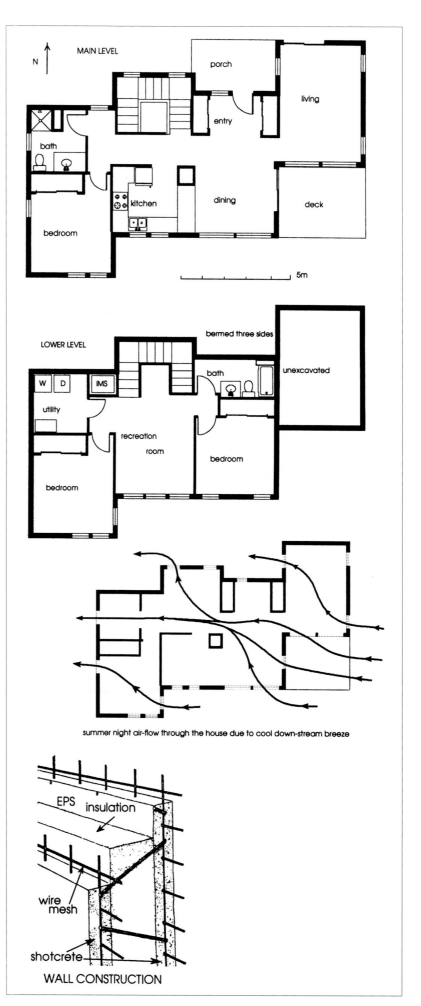

MAIN LEVEL

LOWER LEVEL

summer night air-flow through the house due to cool down-stream breeze

WALL CONSTRUCTION

American Single-family House at Yosemite

Design team:

U.S. National Park Service

OZ Architecture
1580 Lincoln, Suite 200
USA-80203 Denver Co

J. Douglas Balcomb
National Renewable Energy
Laboratory (NREL)
1617 Cole Blvd
USA-80401 Golden CO

Energy features:

High-mass foam-core walls
Night-vent cooling strategy
Direct gain
Integrated mechanical system
Efficient lights and appliances

Energy demand:
(kWh/m^2 a)

Space heating	23
Space cooling	0
Water heating	11
Lights & appliances	19
Fans & pumps	2
Total	55

Heated floor area: 150 m^2

Latitude: 38 °N

Solar Radiation (Glob. Vert. S)
Average Monthly Temperature

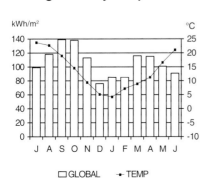

This single-family house, located in El Portal (west of the Yosemite valley) at an altitude of 610 m, must confront a continental climate with hot summers. The average daily temperature swing in July is from 12°C at night to 38°C in the afternoon with extreme peak conditions being still 7°C higher. Cool down-canyon breezes are a natural-cooling resource during summer nights.

The plan incorporates two offsets to enhance ventilation cooling. Air flow from east to west through open windows cools the mass of the house. The enhanced thermal heat capacity of the house allows it to coast comfortably through the hot afternoons when the windows are closed. Sun shading, well-insulated walls and a lower floor bermed on three sides minimise summer solar gains. Ceiling fans provide internal air motion to enhance comfort.

The factory-built walls of 102 mm foam core sandwiched in a wire mesh are sprayed with 51 mm of concrete on both sides. They provide the thermal mass needed for cooling with minimal thermal bridging to the outside. Window solar gains cover much of the winter heat demand. An integrated mechanical system pumping heat from or to exhaust air (COP >3) provides controlled ventilation, all the domestic water heating and back-up cooling, and 56 percent of the space heating.

MONTHLY ENERGY BALANCES

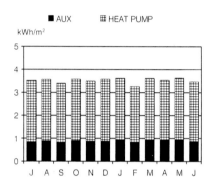

■ AUX ▨ INT ⊞ HEAT REC ▧ SOL (PAS)

kWh/m²

Space Heating

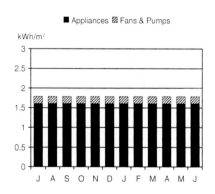

■ AUX ⊞ HEAT PUMP

kWh/m²

Domestic Hot Water

■ Appliances ▨ Fans & Pumps

kWh/m²

Electricity

U-Values: (W/m·K)

Roof	0.13
Wall	0.21
Window	1.83
Door	0.96

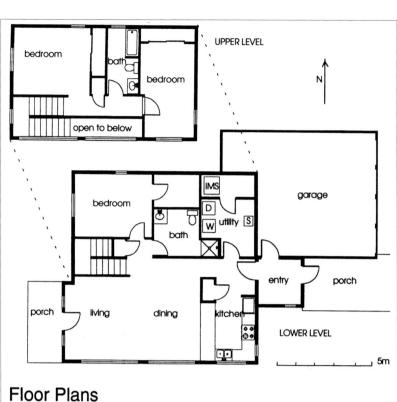

UPPER LEVEL

N

bedroom
bath
bedroom
open to below

IMS
bedroom
D
W utility S
bath
garage

entry
porch
living
dining
kitchen
porch

LOWER LEVEL

5m

Floor Plans

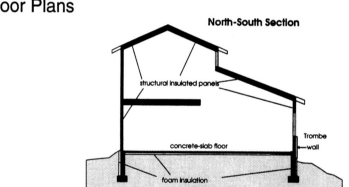

North-South Section

structural insulated panels

concrete-slab floor

Trombe wall

foam insulation

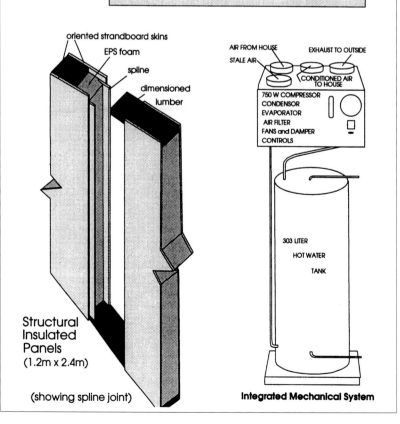

oriented strandboard skins

EPS foam

spline

dimensioned lumber

AIR FROM HOUSE
EXHAUST TO OUTSIDE
STALE AIR
CONDITIONED AIR TO HOUSE

750 W COMPRESSOR
CONDENSOR
EVAPORATOR
AIR FILTER
FANS and DAMPER
CONTROLS

303 LITER
HOT WATER
TANK

Structural Insulated Panels
(1.2m x 2.4m)

(showing spline joint)

Integrated Mechanical System

International Energy Agency

The International Energy Agency, headquartered in Paris, was founded in 1974 as an autonomous body within the framework of the Organization for Economic Cooperation and Development (OECD) to coordinate the energy policies of its members. The twenty-three member countries seek to create the conditions in which the energy sectors of their economies can make the fullest possible contribution to sustainable economic development and the well-being of their people and the environment.

The policy goals in the IEA countries include diversity, efficiency and flexibility within the energy sector, the ability to respond promptly and flexibly to energy emergencies, the environmentally sustainable provision and use of energy, more environmentally acceptable energy sources, improved energy efficiency, research, development and market deployment of new and improved energy technologies, and cooperation among all energy market participants.

These goals are addressed in part through a programme of collaboration in research, development and demonstration of new energy technologies consisting of about 40 Implementing Agreements. The IEA's R&D activities are headed by the Committee on Energy Research and Technology (CERT) which is supported by a small Secretariat staff in Paris. In addition, four Working Parties (in Conservation, Fossil Fuels, Renewable Energy and Fusion) are charged with monitoring the various collaborative agreements, identifying new areas for cooperation and advising the CERT on policy matters.

IEA Solar Heating and Cooling Programme

The Solar Heating and Cooling Programme was one of the first collaborative R&D agreements to be established within the IEA, and, since 1977, its Participants have been conducting a variety of joint projects in passive solar, active solar and photovoltaic technologies, primarily for building applications. The twenty members are:

Australia	France	Spain
Austria	Germany	Sweden
Belgium	Italy	Switzerland
Canada	Japan	Turkey
Denmark	Netherlands	United Kingdom
European Union	New Zealand	United States
Finland	Norway	

A total of 21 projects or "Tasks" have been undertaken since the beginning of the Solar Heating and Cooling Programme. The overall Programme is monitored by an Executive Committee consisting of one representative from each of the member countries. The leadership and management of the individual Tasks are the responsibility of Operating Agents.

These Tasks and their respective Operating Agents are:

*Task 1:   Investigation of the Performance of Solar Heating and Cooling Systems – Denmark
*Task 2:   Coordination of Research and Development on Solar Heating and Cooling – Japan
*Task 3:   Performance Testing of Solar Collectors – Germany/United Kingdom
*Task 4:   Development of an Insulation Handbook and Instrument Package – United States
*Task 5:   Use of Existing Meteorological Information for Solar Energy Application – Sweden
*Task 6:   Solar Systems Using Evacuated Collectors – United States
*Task 7:   Central Solar Heating Plants with Seasonal Storage – Sweden
*Task 8:   Passive and Hybrid Solar Low Energy Buildings – United States
*Task 9:   Solar Radiation and Pyranometry Studies – Canada/Germany
*Task 10: Material Research and Development – Japan
*Task 11: Passive and Hybrid Solar Commercial Buildings – Switzerland
Task 12:   Building Energy Analysis and Design Tools for Solar Applications – United States
Task 13:   Advanced Solar Low Energy Buildings – Norway
Task 14:   Advanced Active Solar Systems – Canada
Task 15:   Not initiated
Task 16:   Photovoltaics in Buildings – Germany
Task 17:   Measuring and Modelling Spectral Radiation – Germany
Task 18:   Advanced Glazing Materials – United Kingdom
Task 19:   Solar Air Systems – Switzerland
Task 20:   Solar Energy in Building Renovation – Sweden
Task 21:   Daylighting in Buildings – Denmark

*Completed

Task 13: Advanced Solar Low Energy Buildings

Energy consumption for space heating has in many countries been greatly reduced over the last few years. This has been mostly achieved by the use of traditional energy conservation and solar technologies. Total energy consumption, especially in residential buildings, is, however, still large and warrants considerable effort. To obtain a significant further reduction in the energy consumption for heating, and also to reduce the consumption for cooling, ventilation and lighting, it has become necessary to develop new building concepts. Such concepts require the use of new materials, components and systems.

Task 13 was started for this purpose. Its official objective is "to advance solar building technologies through the identification, development, and testing of new and innovative concepts which have the potential for eliminating or minimizing the use of purchased energy in residential buildings while maintaining acceptable comfort levels".

The focus of the Task is the application of passive and/or active solar technologies for space heating of single-family and multi-family residential buildings. The use of passive and active solar concepts for cooling, ventilation and lighting is also addressed, as well as advanced energy conservation measures to reduce heating and cooling loads.

Since the emphasis is on innovation and long-range (after the year 2000) cost-effectiveness, the materials, components, concepts and systems considered need not be currently feasible, economical or on the mass market today.

In order to accomplish the above objective, the Participants undertake work in three activities:

A: Development and evaluation of design concepts.
     (Lead country: Germany)
B: Testing of components and data analysis.
     (Lead country: Denmark)
C: Synthesis and documentation of results.
     (Lead country: Switzerland)

Fifteen countries participated in Task 13 which was started in September 1989 and is scheduled to last until September 1996.

Other IEA-SHAC Publications:

Task 13	"Component and System Testing", Bjarne Saxhof (editor). Thermal Insulation Laboratory, DTU, DK-2800 Lyngby

Task 13   "Component and System Testing", Bjarne Saxhof (editor). Thermal Insulation Laboratory, DTU, DK-2800 Lyngby

"Technology Simulation Sets", Bart Poel (editor). Damen Consultants, Stationsplein-West 30, NL-6800 AR Arnhem

"Ways to Zero Energy Houses; Design Principles and Example Projects", Robert Hastings and Anne Grete Hestnes (editors). To be published in 1996.

Task 11   "Passive and Hybrid Solar Commercial Buildings", Andrew Seager and Alan Hildon (editors). Energy Technology Support Unit, Harwell Laboratory, Oxfordshire OX11 0RA, UK

"Passive Solar Commercial and Institutional Buildings – A Sourcebook of Examples and Design Insights", Robert Hastings (editor). John Wiley & Sons Ltd, Baffins Lane, Chichester,   West Sussex PO19 1UD, UK

Task 14   "Low Flow System", William Duff (editor). CSU University, USA – 80523 Fort Collins, CO. (to be published in 1995)

Task 19   "Solar Air Heated Buidings", Andrea Georgi (editor). Forschungsstelle Solararchitektur, ETH-Hönggerberg, CH-8093 Zürich

Task 20   "Solar Energy in Building Renovation", Arne Elmroth and Elisabeth Kjellson (editors). Dept. of Building Physics, Lund University, Lund, Sweden

EXCO   "Program Description and Publications", Sheila Blum (editor). International Planning Assoc., 807 Caddington Ave., Silver Spring, MD 20901, USA

"IEA – Workshop on Thermal Energy Storage and Low Energy Buildings", Wolfgang Schölkopf (editor). W. Schölkopf, Amalienstrasse 54, D-80799 München